5th Grade Math
Volume 7

© 2013 OnBoard Academics, Inc
Newburyport, MA 01950
800-596-3175
www.onboardacademics.com

ISBN: 978-1494857288

Table of Contents

Volume of a Rectangular Prism

Key Vocabulary

volume

rectangular prism

unit cube

How many unit cubes are in a prism?_____

Unit Cube

Do you need to count every cube?

_____

_____

Draw a prism using 27 cubes.

Draw a prism using 16 cubes.

How many cubes fit into this prism?

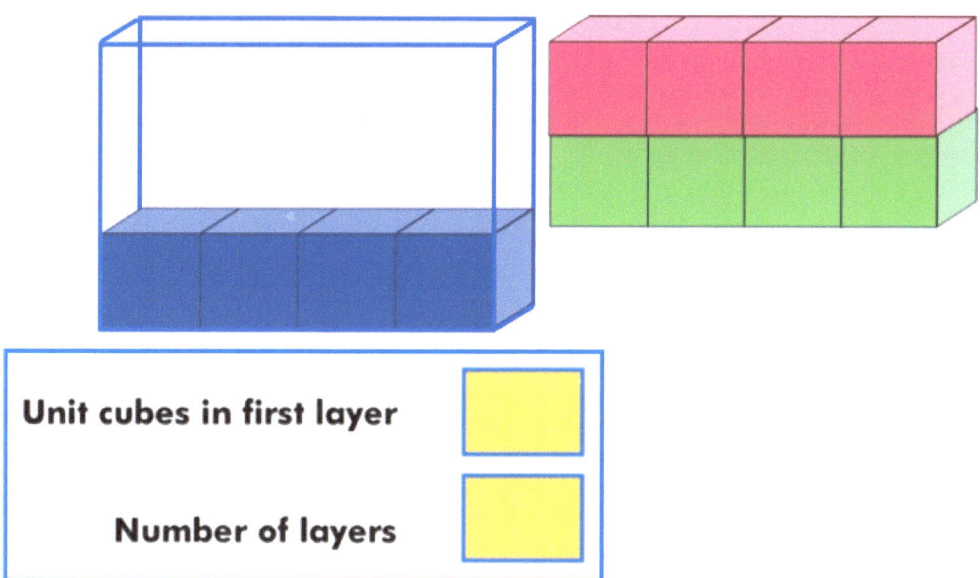

Unit cubes in first layer

Number of layers

How many cubes fit into this prism?

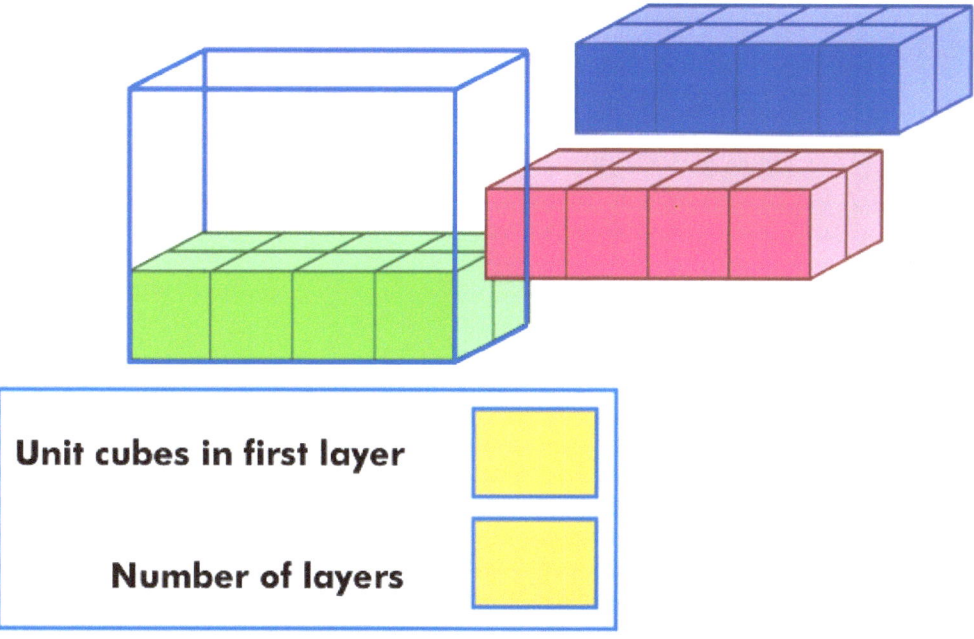

Unit cubes in first layer

Number of layers

How many cubes are there in a prism?

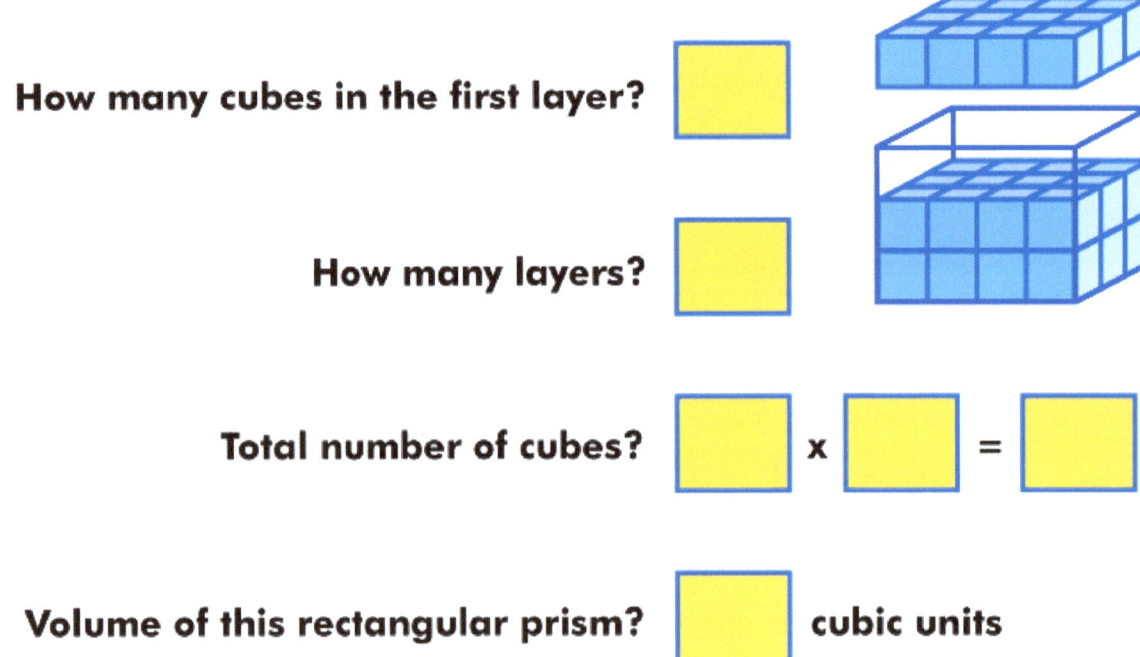

How many cubes in the first layer? ▢

How many layers? ▢

Total number of cubes? ▢ x ▢ = ▢

Volume of this rectangular prism? ▢ cubic units

Find the volumes of these prisms.

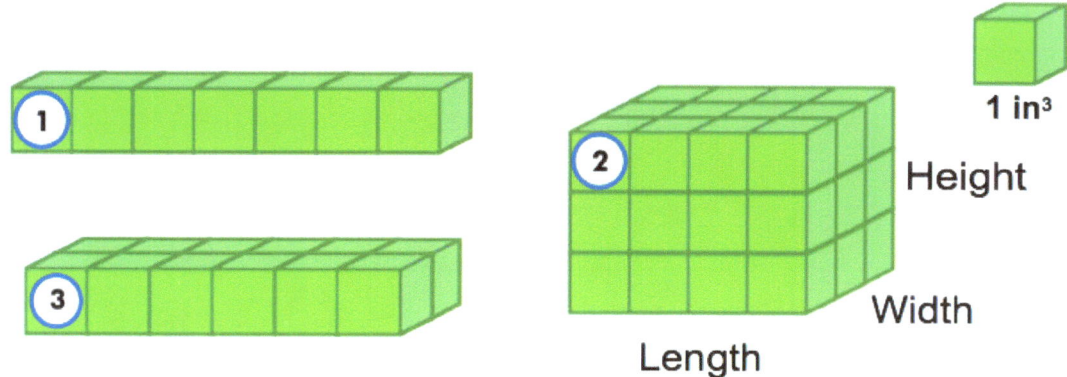

	Length	Width	Height	Volume
1	in	in	in	in³
2	in	in	in	in³
3	in	in	in	in³

Find the volume of these prisms.

1) _____

2) _____

3) _____

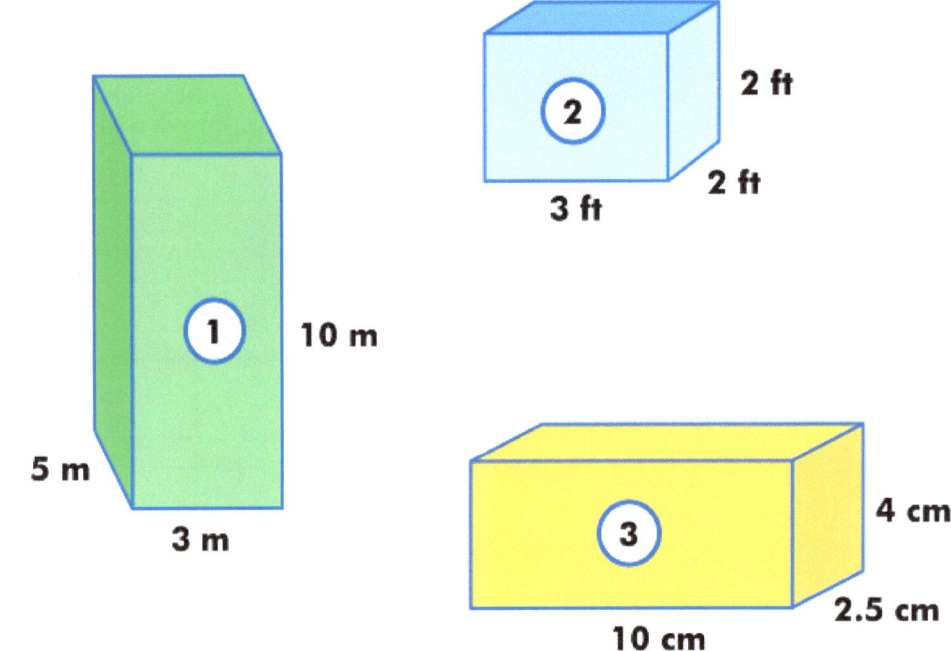

Find the missing dimension.

 **Alison's parents have purchased a swim training pool which holds 6,000 ft³ of water.**

What is the depth of the pool?

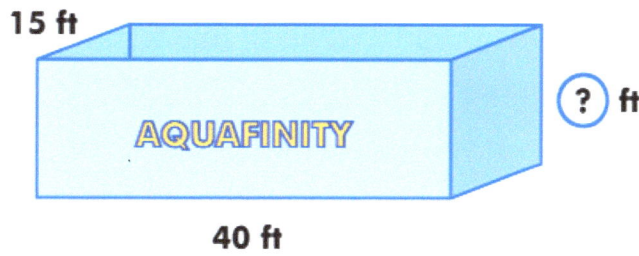

Hint

$$40 \times 15 \times \boxed{?} = 6,000$$

Name: _____

Volume of a Rectangular Prism Quiz

1. True of false? Perimeter, area and volume can all be measured in units of square feet and square inches.
2. What is the formula for the volume of a rectangular prism?
   a. $(l \times w \, h)^3$
   b. l x w x h
   c. 2l x 2w x 2h
   d. l + w + h
3. What is the volume, in cubic units of this figure? _____

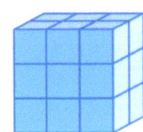

4. What is the volume, in cubic units, of this figure? _____

5. The volume of this prism is 24 units.

   True

   False

Solid Figures

Key Vocabulary

solid figure

base

vertex

edge

face

net

Draw a line to connect the shape with its name.

Sphere   Rectangular Prism   Cube

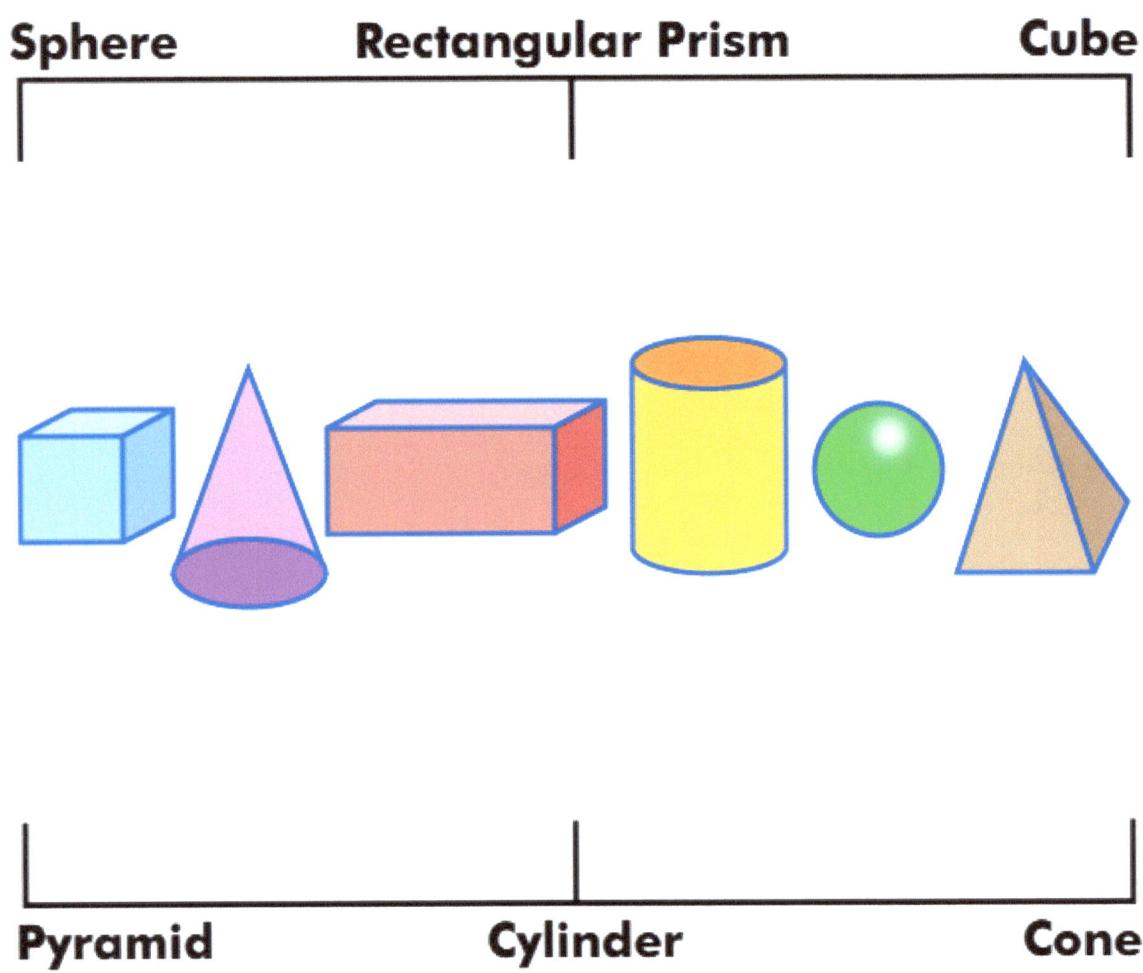

Pyramid   Cylinder   Cone

Prisms

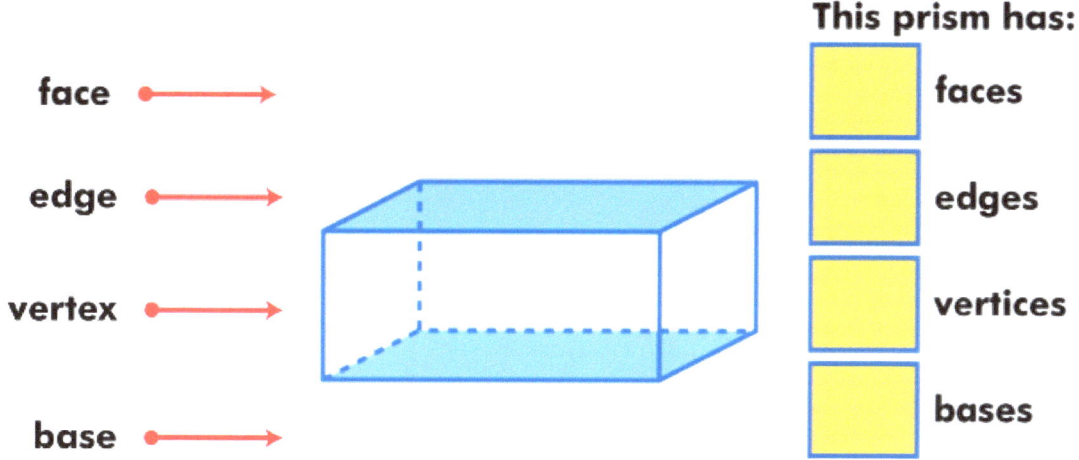

This prism has:

	faces
	edges
	vertices
	bases

face ●——→

edge ●——→

vertex ●——→

base ●——→

Label one face, one vertex, one edge, and one base of this prism.

A prism has two parallel congruent bases. It is named by the shape of its base. The other faces are always rectangles.

Pyramids

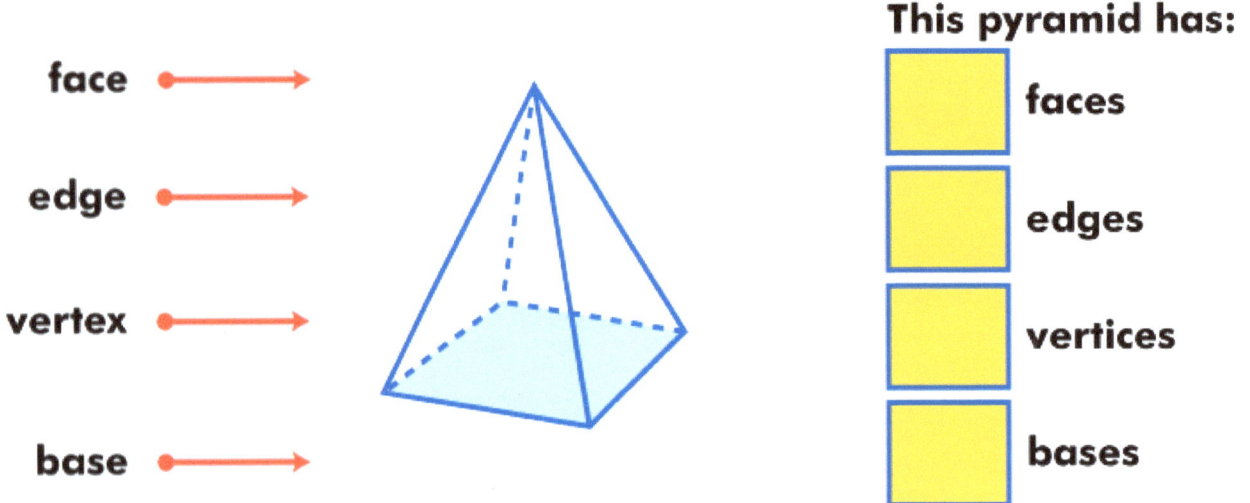

face ⬤➞

edge ⬤➞

vertex ⬤➞

base ⬤➞

This pyramid has:

faces

edges

vertices

bases

Label one face, one vertex, one edge, and the base of this pyramid.

**A pyramid has one base. It is named by the shape of its base.
The other faces are always triangles.**

Three solids with curved surfaces

Label the base(s) and curved surface(s) of each solid

Cylinder

Cone

Sphere

curved surface

base

Label the illustrations below

| Prism | or | Pyramid |

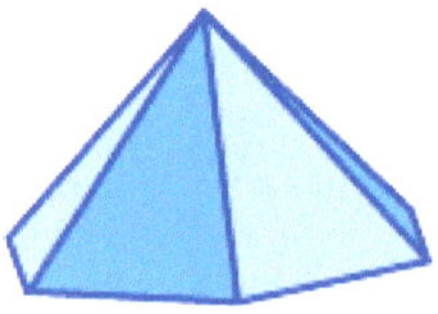

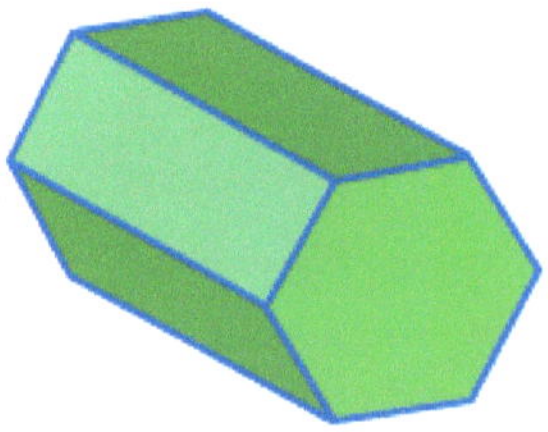

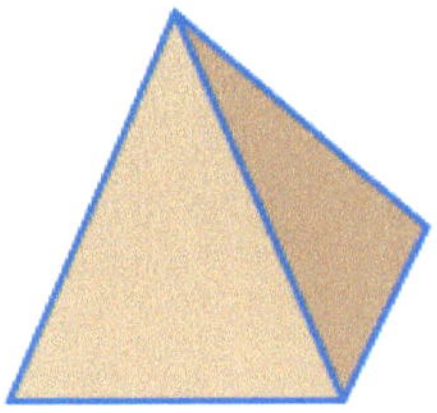

One base

Two parallel bases

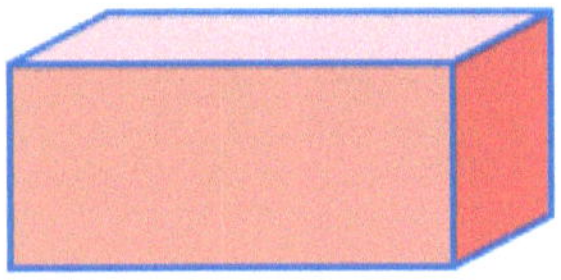

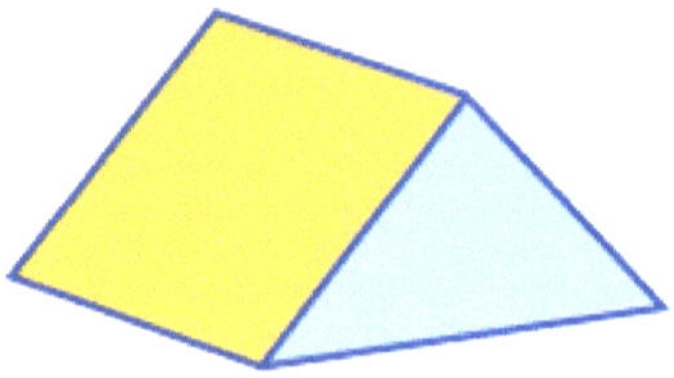

The unfolded box

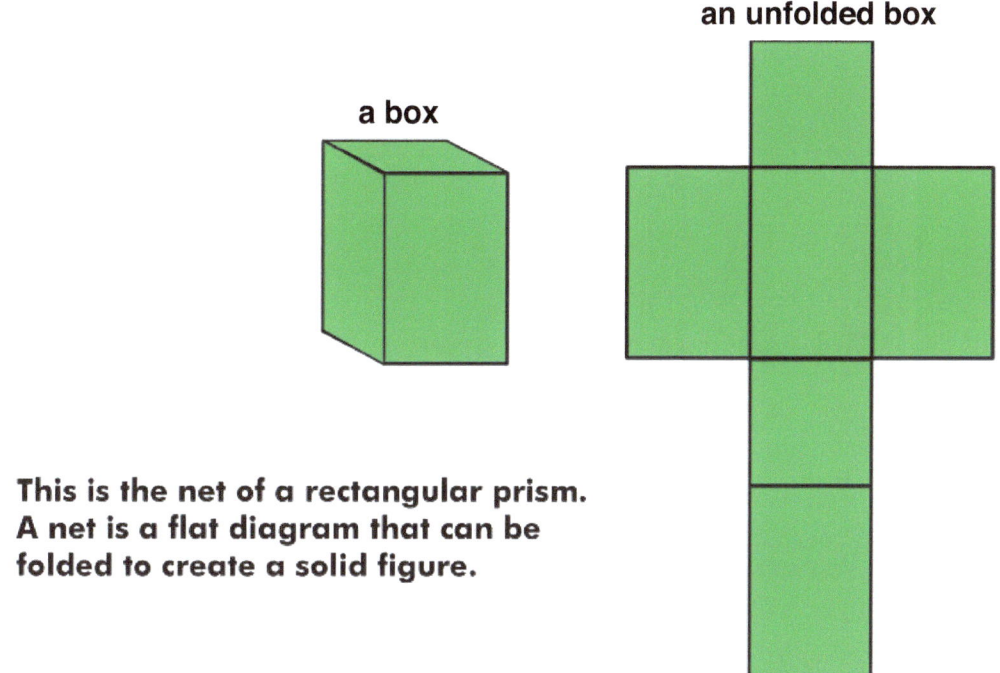

a box

an unfolded box

This is the net of a rectangular prism. A net is a flat diagram that can be folded to create a solid figure.

Match the solid with its unfolded version.
Enter the number of the yellow unfolded solid in the box below the corresponding blue solid.

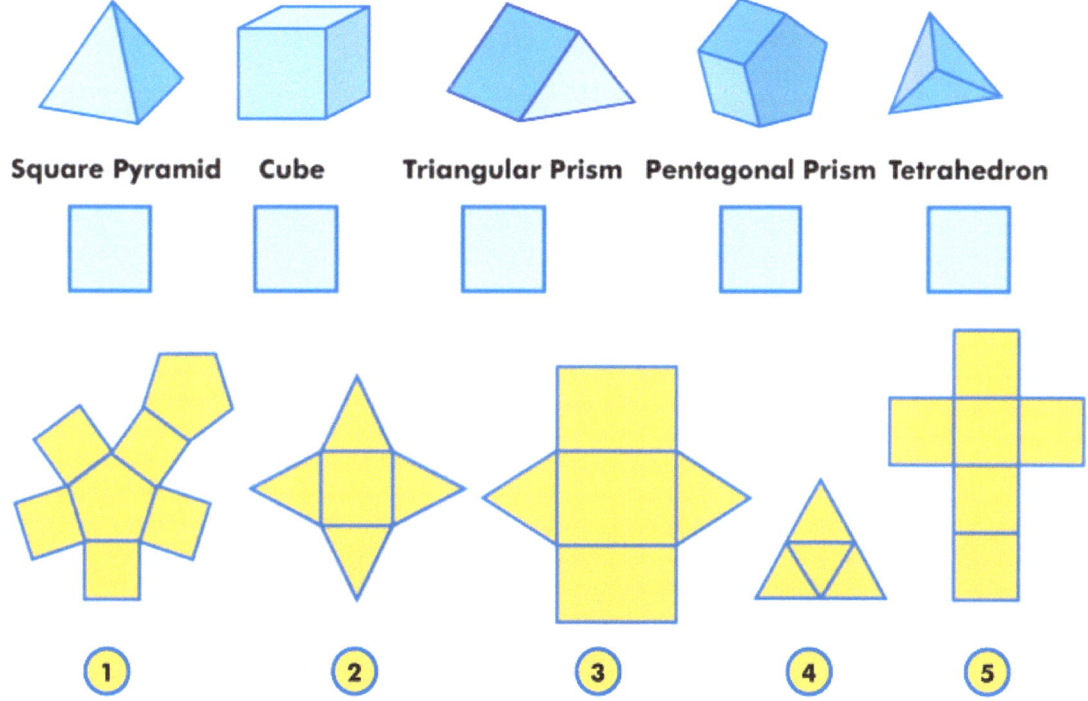

Square Pyramid   **Cube**   **Triangular Prism**   **Pentagonal Prism**   **Tetrahedron**

Name: _____

Solid Figures Quiz

1 True or false? A prism has two parallel congruent bases. The other faces are always triangles.

2 This solid has five faces, five vertices, and eight edges. Four of the faces are triangles.

- **A** Rectangular prism
- **B** Triangular prism
- **C** Square pyramid
- **D** Pentagonal pyramid

3 How many faces does a triangular prism have?

4 How many edges does a cube have?

Measures of Central Tendency

Key Vocabulary

central tendency

mean

median

mode

range

line plot

Backpack weight survey

Students in Anthony's class weighed their backpacks.
Order the results below (rounded to the nearest lb).

| 9 | 9 | 10 | 12 | 8 | 9 | 10 | 11 | 4 | 11 | 8 | 13 | 16 |

LEAST ————————————————————————→ GREATEST

Complete the line plot for this data.

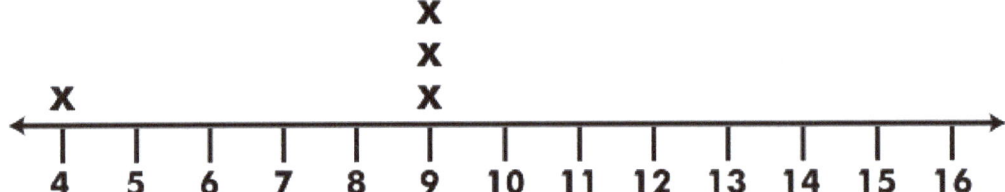

Weight of student backpacks (lb)

```
                              X
                              X
     X                        X
  ←——┬——┬——┬——┬——┬——┬——┬——┬——┬——┬——┬——┬——→
     4  5  6  7  8  9  10 11 12 13 14 15 16
```

Describe the distribution of the data.

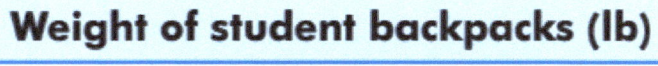

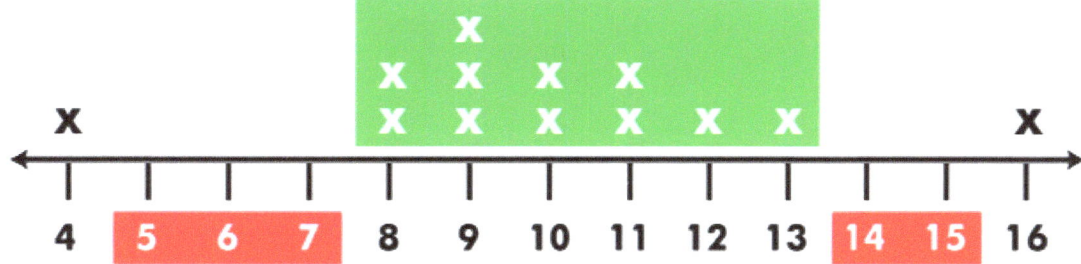

Are there any gaps?

What does a gap mean?

Are there any clusters?

What do clusters mean?

Mode and Median

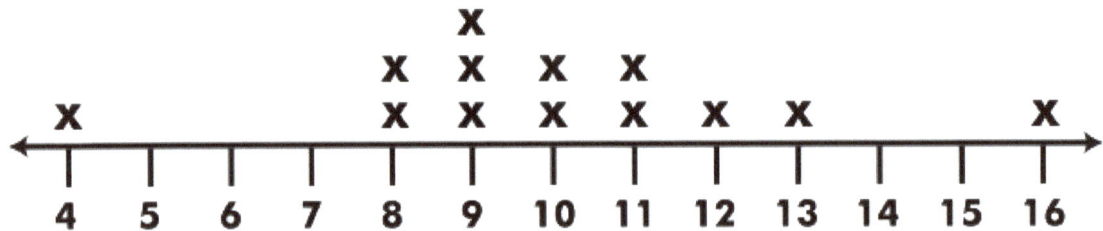

Find the mode.

The mode is the tallest column of X's. It is the most common number.

Find the median.

There are 13 values and so the median is the 7th value. It is the middle value; there are six values before and six after.

Range

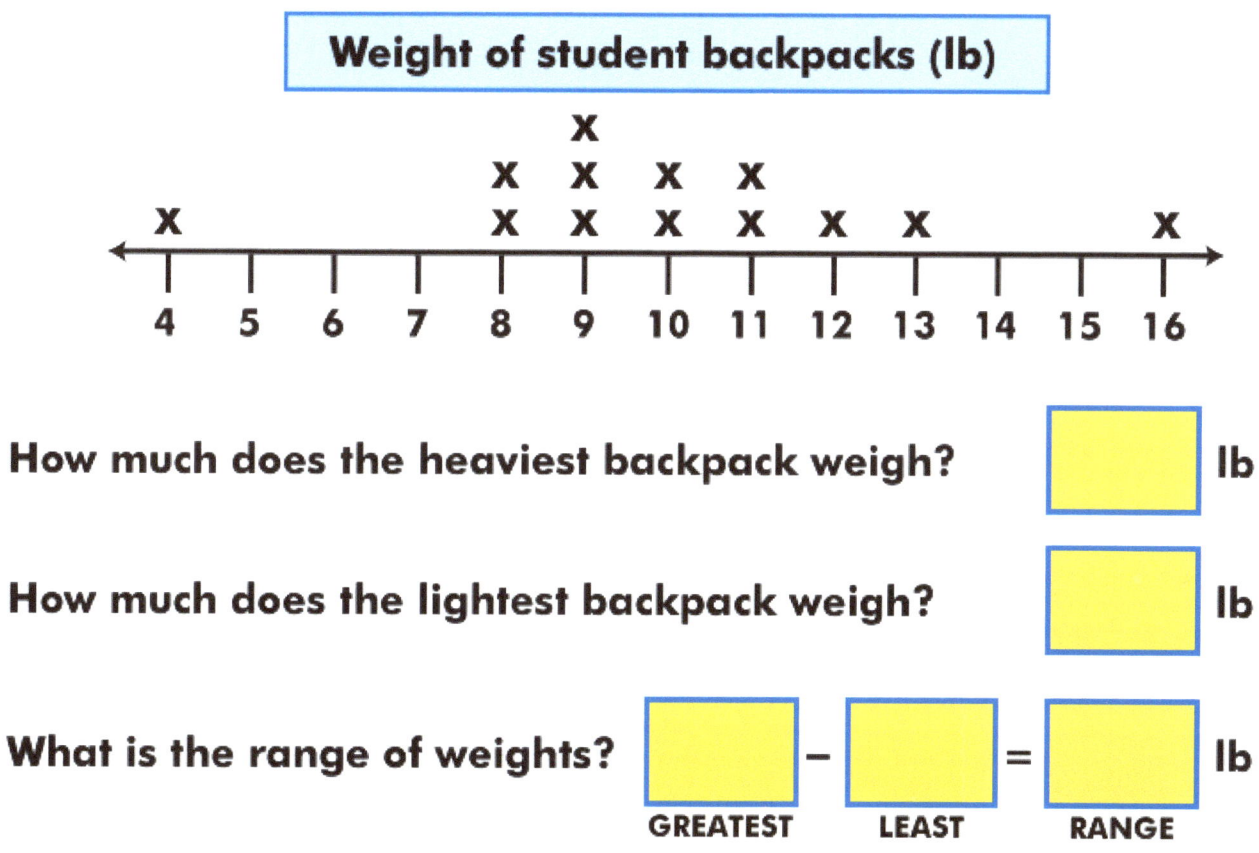

How much does the heaviest backpack weigh? [ ] lb

How much does the lightest backpack weigh? [ ] lb

What is the range of weights? [ ] – [ ] = [ ] lb
GREATEST   LEAST   RANGE

Mean

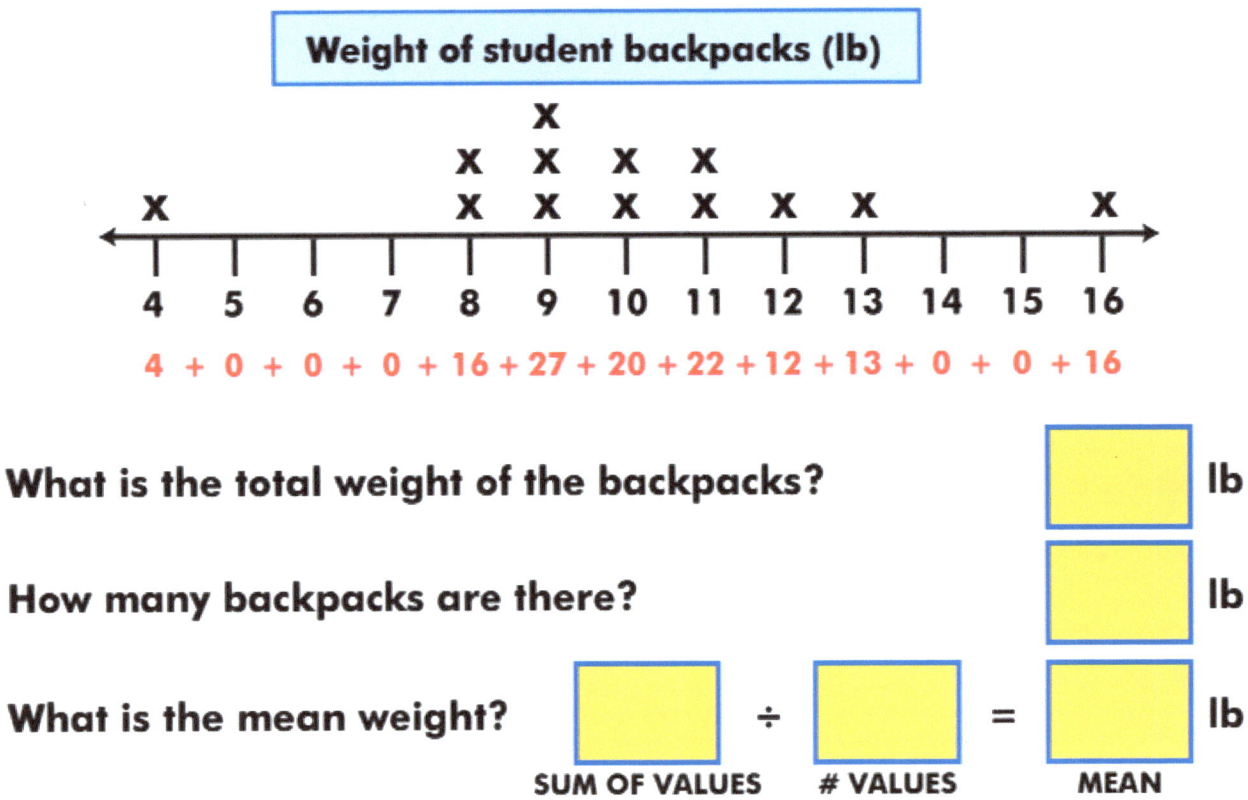

Weight of student backpacks (lb)

4 + 0 + 0 + 0 + 16 + 27 + 20 + 22 + 12 + 13 + 0 + 0 + 16

What is the total weight of the backpacks?   ⬜ lb

How many backpacks are there?   ⬜ lb

What is the mean weight?   ⬜ ÷ ⬜ = ⬜ lb
                          SUM OF VALUES   # VALUES   MEAN

Find the mean, mode, median and range.

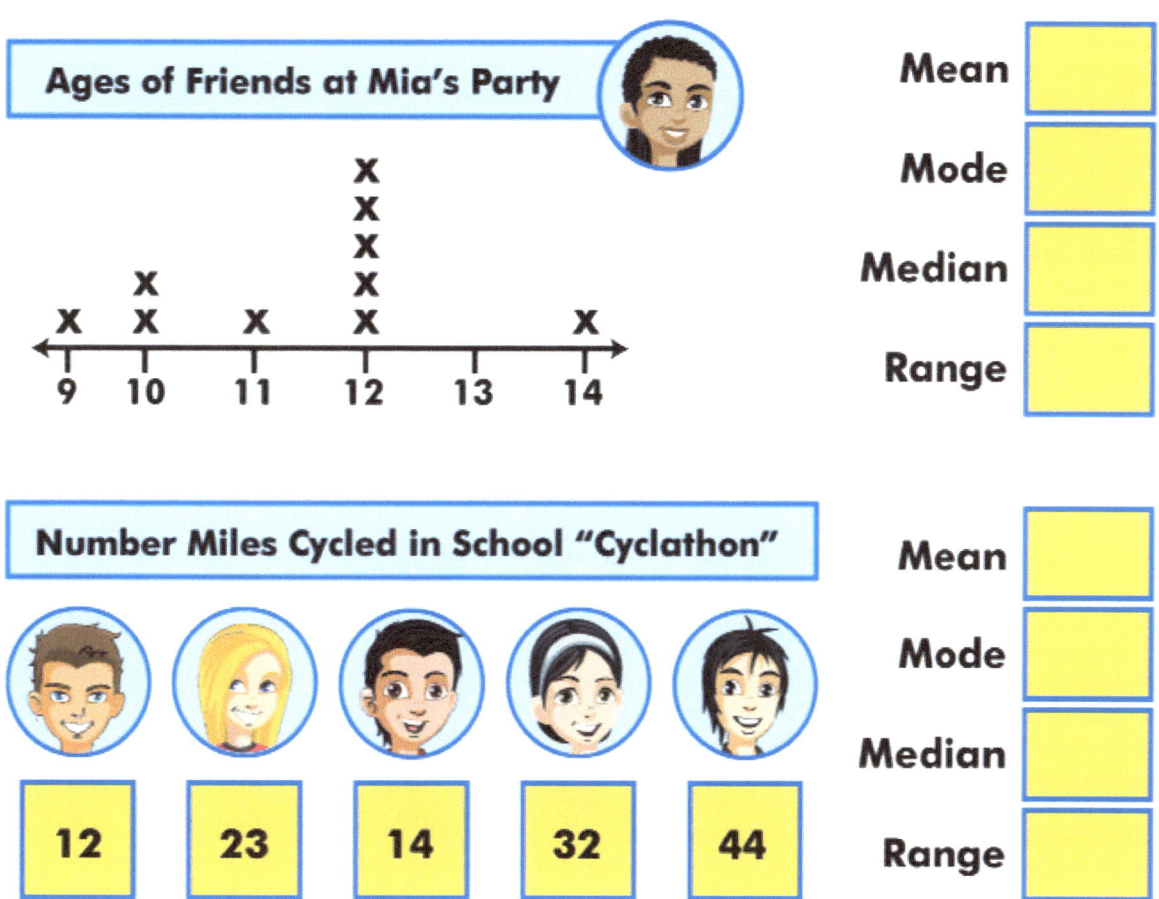

How many nuggets on Friday?

 **Each time that Jack purchases McCluck Nuggets, he records the number of nuggets in a box.**

Mon	Tues	Weds	Thurs	Fri	Sat	Sun
13	11	12	11	?	12	12

The mean number of nuggets in a box is 12.

How many nuggets were in the box on Friday?

Fri

Name: _____

Measures of Central Tendency Quiz

1   True or false? A set of data can have more than one mode.

2   Which statement is *not* correct?

   A   The mode is the value that occurs most often

   B   The median is the middle value

   C   The mean is the sum of all values

   D   The range is the greatest value minus the least value

3   Find the median value: {15, 12, 13, 14, 11, 14}

4   The mean of {12, 14, n, 14, 9} is 12. Find the value of n.

Displaying Data

Key Vocabulary

bar graph

stem-and-leaf plot

mean

median

mode

range

Two ways to display data
Discover two ways to display data below.

Data

Class 5C Favorite Ice Cream

Flavor	# Students
Chocolate/Chip	卌 II
Strawberry	卌 卌
Vanilla	卌
Other	III

Bar Graph

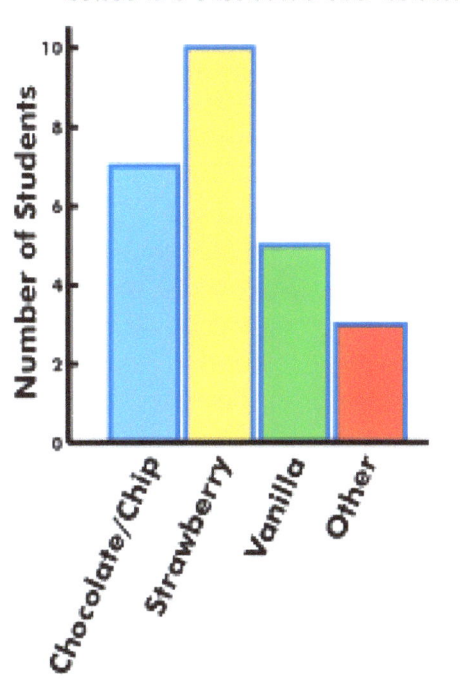

Data

Class 5D 10K Road Race Times				
54	85	78	77	65
67	51	73	64	66

Times rounded to nearest minute.

Stem and Leaf

Class 5D 10K Road Race Times

Stem	Leaf
5	1 4
6	4 5 6 7
7	3 7 8
8	5

Key

5 | 1 = 51

Times rounded to nearest minute.

Label the bar graph.
Labels are provided below.

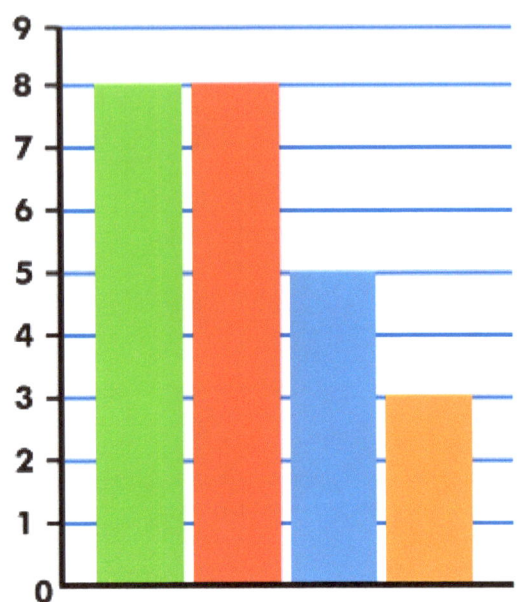

6B Favorite Sports Survey

Sport	# Students
Baseball	5
Basketball	8
Football	8
Tennis	3

Number of students

Favorite sports

Football Baseball Basketball Tennis

Survey of Favorite Sports – Class 6B

Use the bar graph to estimate the fund raising totals.

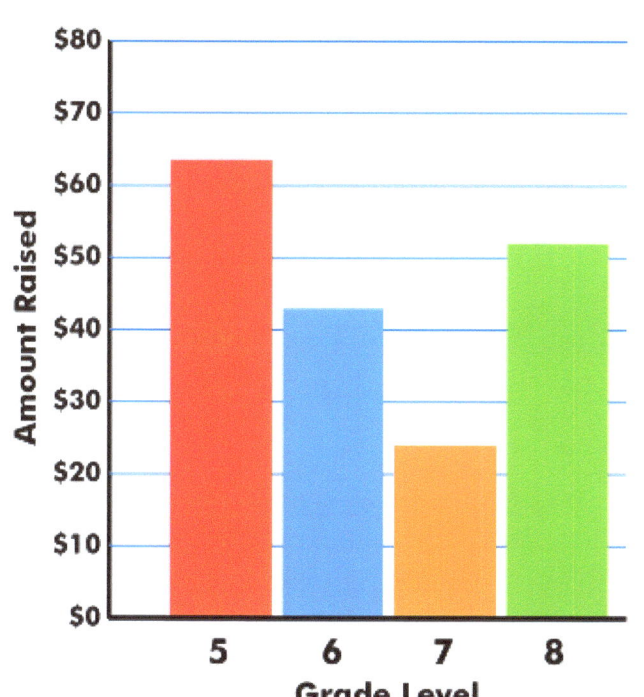

'Walk for Charity' Fundraiser

Amount Raised by Grade

Grade	Amount
5	$
6	$
7	$
8	$

Estimate what fraction of the total amount was raised by Grade 5?

Which individual student raised the most money?

Complete the tables for the road race times.
Race times are rounded to the nearest minute.

54   85   78   77   65   67   51   73   64

5-Mile Road Race Times

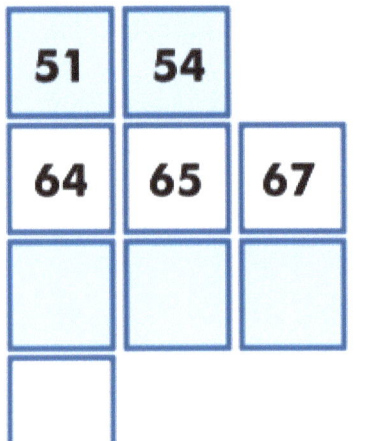

51	54	
64	65	67

5-Mile Road Race Times

Stem	Leaf		
5	1	4	
6			

Stem and Leaf Plot

5-Mile Road Race Times

Stem	Leaf
5	1 4
6	4 5 7
7	3 7 8
8	5

Key 5|4 = 54

Amy took 105 minutes to complete the race. Add her time to the stem-and-leaf plot. Describe the distribution of the data.

_____

_____

_____

_____

_____

Complete the stem and leaf plot.

 Points scored by MVP during playoffs:
22, 23, 31, 32, 32, 34

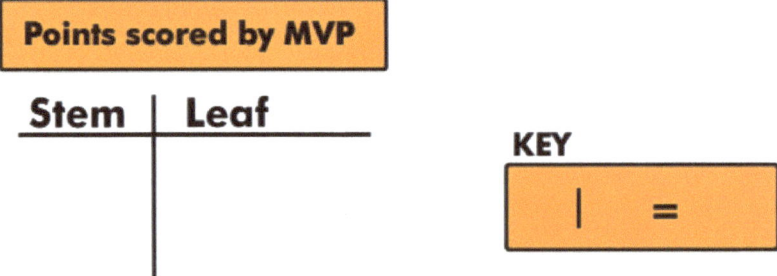

Study the illustration below to discover how to use the stem and leaf plot to find mean, median and mode.

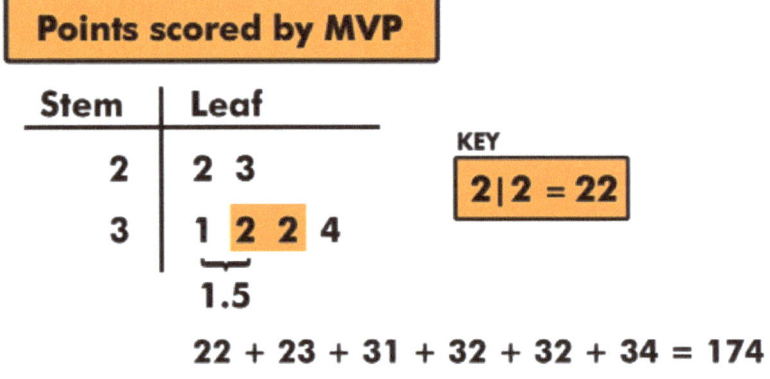

22 + 23 + 31 + 32 + 32 + 34 = 174

The mean is 29 $\dfrac{174}{6} = 29$

The median is 31.5

The mode is 32

Name: _____

Displaying Data Quiz

1 It's not always necessary to label the axes on a bar graph.

2 What is the mode of this data?

A 14

B 21

C 14 and 21

D 11 and 14

5B Weekly TV Hours

Stem	Leaf
1	1 4 4 5 6 9
2	1 1 7
3	2 5
4	2

KEY

2|7 = 27

3 How many students took part in this survey?

4 What fraction of the class watch more than 25 hours of TV per week?

Probability

Key Vocabulary

probability

event

certain

impossible

theoretical probability

experimental probability

What is the probability?

Consider some "in the future" statements and place them on the probability line. Write the number of the "in the future" statement on the number line to indicate probability.

1. Tonight I will watch TV.
2. Tomorrow I will eat lunch.
3. I will play football for the New England Patriots when I am 25.
4. My hair will fall out when I'm older.
5. I will have a pet pig when I am older.

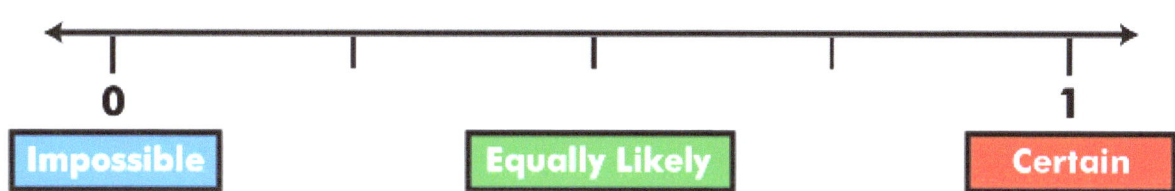

If you flip a coin, what is the probability of any of these events happening.
Place the number on the probability line.

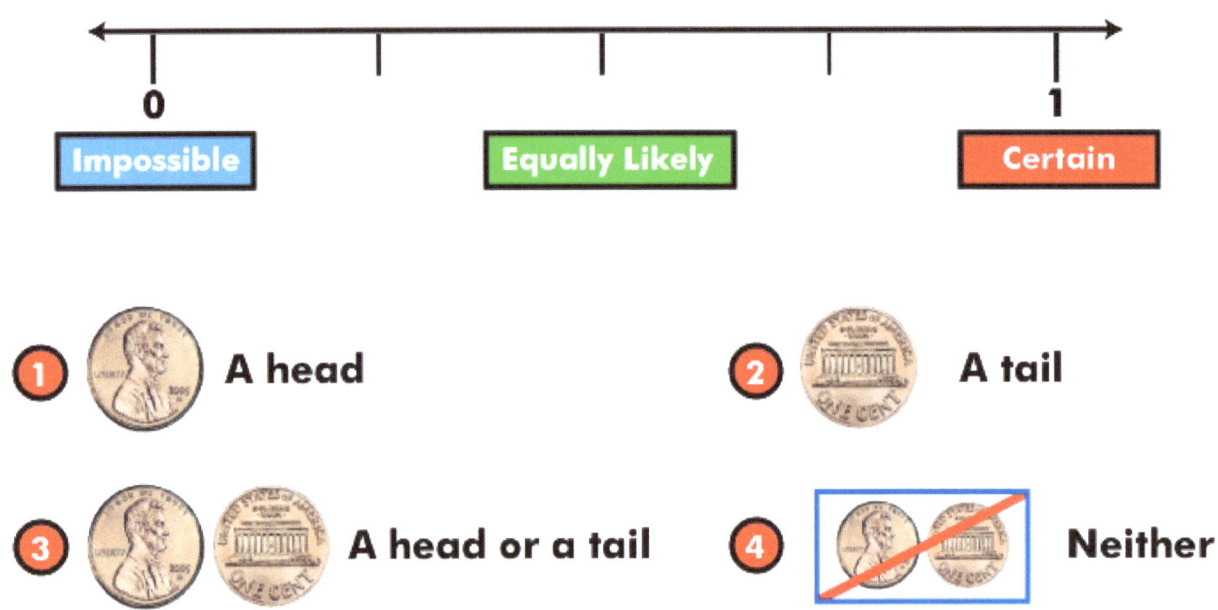

What is the probability?

$$p(event) = \frac{\text{\# favorable outcomes}}{\text{\# possible outcomes}}$$

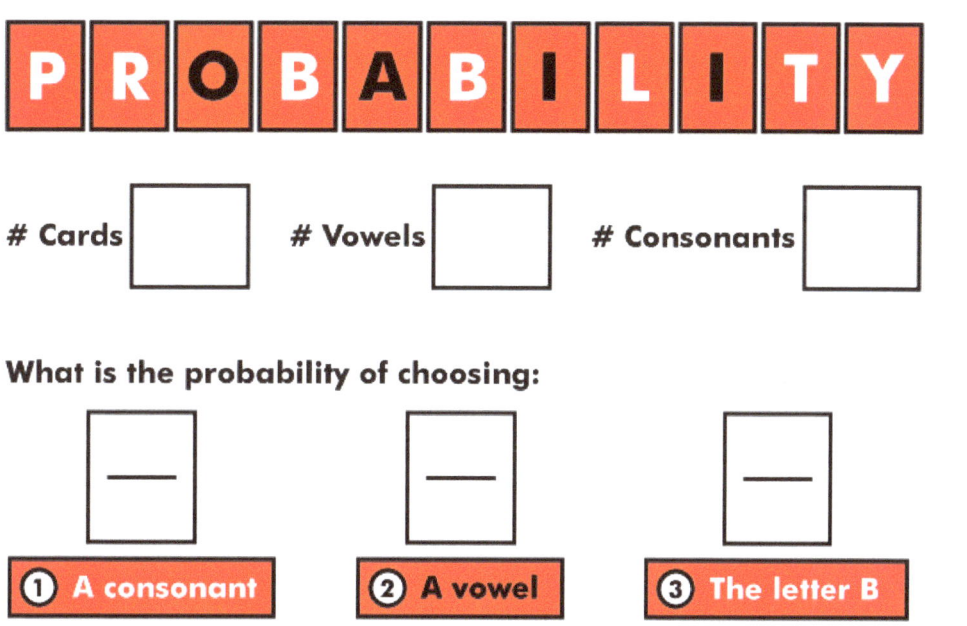

Cards [   ]   # Vowels [   ]   # Consonants [   ]

What is the probability of choosing:

① A consonant   ② A vowel   ③ The letter B

_____

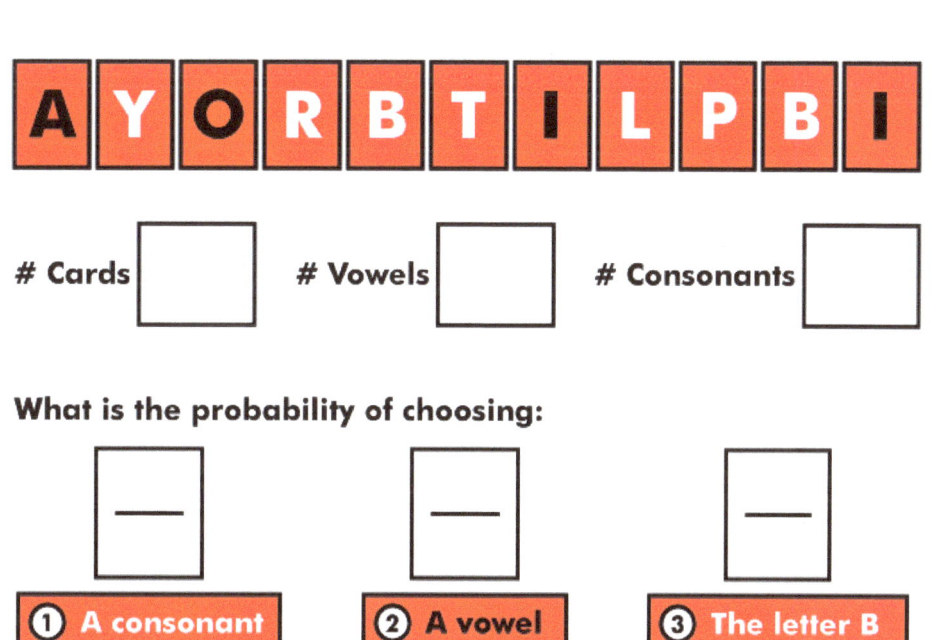

Cards [   ]   # Vowels [   ]   # Consonants [   ]

What is the probability of choosing:

① A consonant   ② A vowel   ③ The letter B

For each spinner, what is the probability of spinning blue?
Draw a line from the spinner to the correct place on the probability line.

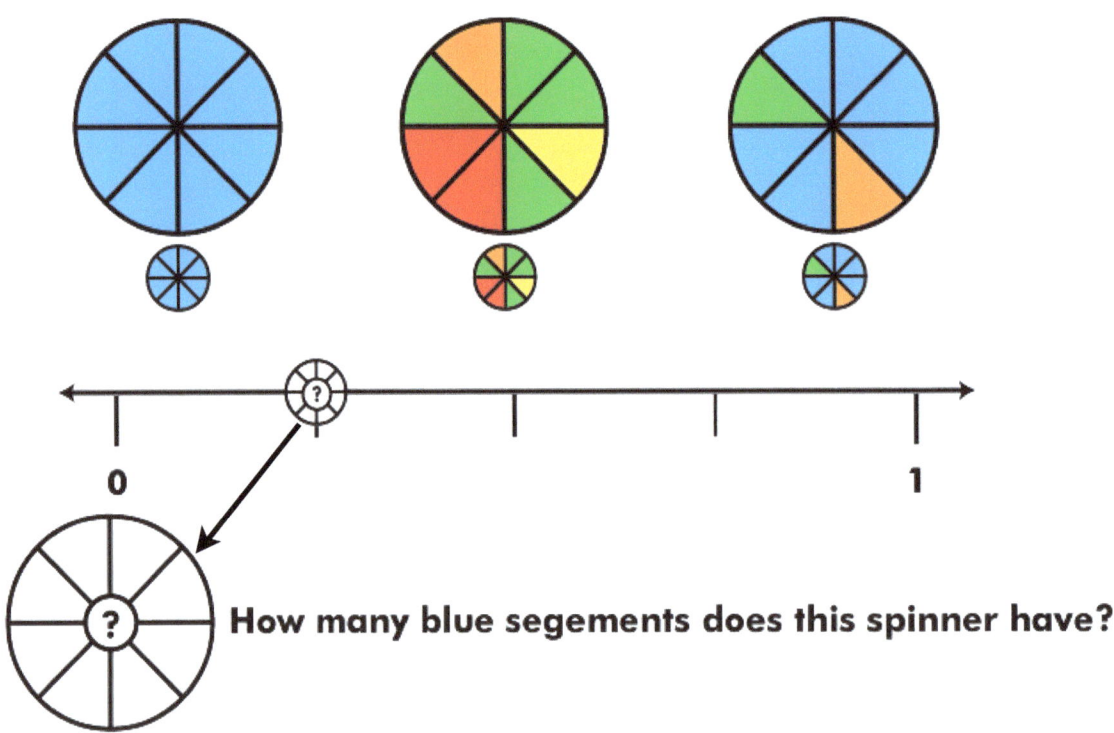

How many blue segements does this spinner have?

What is the probability of picking a red ball?
Shade the balls blue and red or simply use the letter R and B to indicate color.

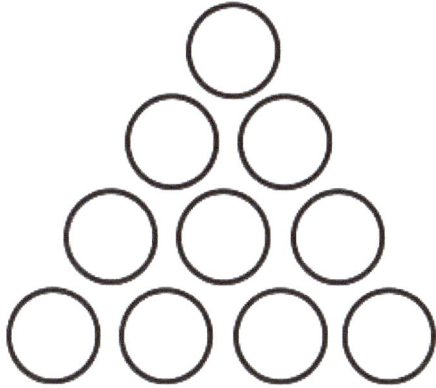

The probability of picking a blue ball is $\dfrac{3}{10}$

The probability of picking a green ball is $\dfrac{1}{5}$

If the rest of the balls are red, what is the probability of picking a red ball? ⬜ **or** ⬜

Spinners

Total number of sectors [ ]

What is the probability of the spinner landing on:

Green [———] or [———]

Red [———]

Purple [———] or [———]

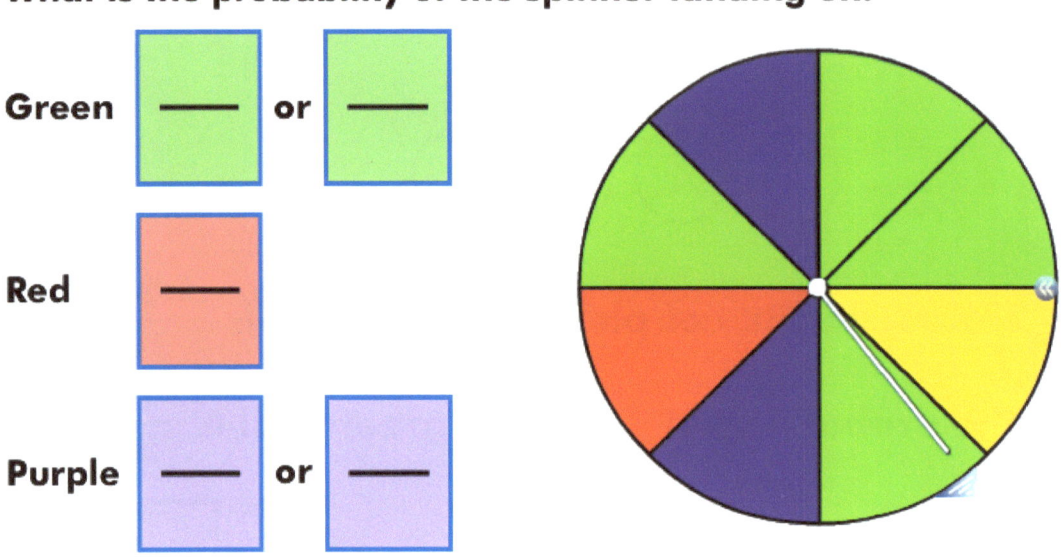

$$p(green) = \frac{\text{\# green sectors}}{\text{total \# sectors}}$$

Name: _____

Probability Quiz

1 True or false? The probability of randomly picking a green ball is $\frac{1}{2}$.

2 What is the probability of randomly picking a blue ball?

A $\frac{2}{5}$

B $\frac{3}{4}$

C $\frac{4}{5}$

D $\frac{5}{10}$

3 What is the probability of randomly picking a yellow ball?

4 What is the probability of randomly picking a red, green, or blue ball?

Newburyport, MA 01950

1-800-596-3175

OnBoard Academics employs teachers to make lessons for teachers! We create and publish a wide range of aligned lessons in math, science and ELA for use on most EdTech devices including whiteboard, tablets, computers and pdfs for printing.

All of our lessons are aligned to the common core, the Next Generation Science Standards and all state standards.

If you like our products please visit our website for information on individual lessons, teachers licenses, building licenses, district licenses and subscriptions.

Thank you for using OnBoard Academic products.